APPLICATIONS OF GRAPH THEORY

ASHAY DHARWADKER

SHARIEFUDDIN PIRZADA

COPYRIGHT © 2007
INSTITUTE OF MATHEMATICS
H-501 PALAM VIHAR, GURGAON, HARYANA 122017, INDIA
www.dharwadker.org

Abstract

Graph theory is becoming increasingly significant as it is applied to other areas of mathematics, science and technology. It is being actively used in fields as varied as biochemistry (genomics), electrical engineering (communication networks and coding theory), computer science (algorithms and computation) and operations research (scheduling). The powerful combinatorial methods found in graph theory have also been used to prove fundamental results in other areas of pure mathematics. This paper, besides giving a general outlook of these facts, includes new graph theoretical proofs of Fermat's Little Theorem and the Nielson-Schreier Theorem. New applications to DNA sequencing (the SNP assembly problem) and computer network security (worm propagation) using minimum vertex covers in graphs are discussed. We also show how to apply edge coloring and matching in graphs for scheduling (the timetabling problem) and vertex coloring in graphs for map coloring and the assignment of frequencies in GSM mobile phone networks. Finally, we revisit the classical problem of finding re-entrant knight's tours on a chessboard using Hamiltonian circuits in graphs.

Contents

Introduction

Graph theory is rapidly moving into the mainstream of mathematics mainly because of its applications in diverse fields which include biochemistry, electrical engineering (communications networks and coding theory), computer science (algorithms and computations) and operations research (scheduling). The wide scope of these and other applications has been well-documented cf. [5] [19]. The powerful combinatorial methods found in graph theory have also been used to prove significant and well-known results in a variety of areas in mathematics itself. The best known of these methods are related to a part of graph theory called matchings, and the results from this area are used to prove Dilworth's chain decomposition theorem for finite partially ordered sets. An application of matching in graph theory shows that there is a common set of left and right coset representatives of a subgroup in a finite group. This result played an important role in Dharwadker's 2000 proof of the four-color theorem [8] [18]. The existence of matchings in certain infinite bipartite graphs played an important role in Laczkovich's affirmative answer to Tarski's 1925 problem of whether a circle is piecewise congruent to a square. The proof of the existence of a subset of the real numbers **R** that is non-measurable in the Lebesgue sense is due to Thomas [21]. Surprisingly, this theorem can be proved using only discrete mathematics (bipartite graphs). There are many such examples of applications of graph theory to other parts of mathematics, but they remain scattered in the literature [3] [16]. In this paper, we present a few selected applications of graph theory to other parts of mathematics and to various other fields in general.

1. The Cantor-Schröder-Bernstein Theorem

Here we discuss the graph theoretical proof of the classical result of Schröder and Bernstein. This theorem was presumed to be an obvious fact by Cantor (cf. remark 1.2) and later proved independently by Schröder (1896) and Bernstein (1905). The proof given here can be found in [14] and is attributed to König.

1.1. Theorem (Cantor-Schroder-Bernstein). For the sets A and B, if there is an injective mapping $f: A \rightarrow B$ and an injective mapping $g: B \rightarrow A$, then there is a bijection from A onto B, that is, A and B have the same cardinality.

Proof. Without loss of generality, assume A and B to be disjoint. Define a bipartite graph $G = (A, B, E)$, where $xy \in E$ if and only if either $f(x) = y$ or $g(y) = x$, $x \in A$, $y \in B$. By the hypothesis, $1 \leq d(v) \leq 2$ for each v of G. Therefore, each component of G is either a one-way infinite path (that is, a path of the form $x_0, x_1, \ldots, x_n, \ldots$), or a two-way infinite path (of the form $\ldots, x_{-n}, x_{-n+1}, \ldots, x_{-1}, x_0, x_1, \ldots, x_n, \ldots$), or a cycle of even length with more than two vertices, or an edge. Note that a finite path of length greater or equal to two cannot be a component of G. Thus, in each component there is a set of edges such that each vertex in the component is incident with precisely one of these edges. Hence, in each component, the subset of vertices from A is of the same cardinality as the subset of vertices from B. ■

1.2. Remark. Cantor inferred the result as a corollary of the well-ordering principle. The above argument shows that the result can be proved without using the axiom of choice.

2. Fermat's (Little) Theorem

There are many proofs of Fermat's Little Theorem. The first known proof was communicated by Euler in his letter of March 6, 1742 to Goldbach. The idea of the graph theoretic proof given below can be found in [12] where this method, together with some number theoretic results, was used to prove Euler's generalization to non-prime modulus.

2.1. Theorem (Fermat). Let a be a natural number and let p be a prime such that a is not divisible by p. Then, $a^p - a$ is divisible by p.

Proof. Consider the graph $G = (V, E)$, where the vertex set V is the set of all sequences $(a_1, a_2, \ldots, a_p)$ of natural numbers between 1 and a (inclusive), with $a_i \neq a_j$ for some $i \neq j$. Clearly, V has $a^p - a$ elements. Let $u = (u_1, u_2, \ldots, u_p)$, $v = (u_p, u_1, \ldots, u_{p-1}) \in V$. Then, we say $uv \in E$. With this assumption, each vertex of G is of degree 2. So, each component of G is a cycle of length p. Therefore, the number of components is $\dfrac{a^p - a}{p}$. That is, $p \big/ a^p - a$. ∎

3. The Nielson-Schreier Theorem

Let H be a group and S be a set of generators of H. The product of generators and their inverses which equals identity (1) is called a *trivial relation among the generators in S* if 1 can be obtained from that product by repeatedly replacing xx^{-1} or $x^{-1}x$ by 1. Otherwise such a product is called a *non-trivial relation*. A group H is *free* if H has a set of generators such that all relations among the generators are trivial.

Babai [2] proved the Nielson-Schreier Theorem for subgroups of free groups, as well as other results in diverse areas, from his Contraction Lemma. The particular case of this lemma when G is a tree, and its use in proving the Nielson-Schreier Theorem, was also observed by Serre [20].

3.1. Contraction Lemma. Let H be a semi-regular subgroup of the automorphism group of a connected graph G. Then, G is contractible onto some Cayley graph of H. The proof of this lemma is technical, although it only uses ideas from group theory and graph theory.

Let H be a group and $h \in H$. Let h_R be a permutation of H obtained by multiplying all the elements of H on the right by h. The collection $H_R = \{h_R: h \in H\}$ is a regular group of permutations (under composition) and is called the (right) regular permutation representation of H.

It can be seen [2] that G is a Cayley graph of the group H if and only if G is connected and H_R is a subgroup of the automorphism group of G. The automorphism group of a graph G is the group of all permutations p of the vertices of G with the property that $p(x)p(y)$ is an edge of G if and only if xy is an edge of G.

A group H of permutations acting on a set V is called *semi-regular* if for each $x \in V$, the stabilizer $H_x = \{h \in H: x^h = x\}$ consists of the identity only, where x^h denotes the image of x under h. If H is transitive and semi-regular, then it is regular.

Let (H, o) be a group and S be a set of generators of H, not necessarily minimal. The Cayley graph $G(H, S)$ of (H, o) with respect to S, has vertices $x, y,\ldots \in H$, and xy is an edge if and only if either $x = yoa$ or $y = xoa$, for some $a \in S$.

If G is any graph and $e = xy$ an edge of G, then by contraction along e, we mean the graph G' obtained by identifying the vertices x and y. We say that a graph G_1 is contractible onto a graph G_2 if there is a sequence of contractions along edges which transforms G_1 to G_2.

3.2. Corollary. If J is a subgroup of a group H, then any $G(H, S)$ is contractible onto $G(J, T)$ for some set T of generators of J.

3.3. Theorem (Nielson-Schreier). Any subgroup of a free group is free.

Proof. We first show that in any group H and for any set S of generators of H, the Cayley graph $G(H, S)$ contains a cycle of length > 2 if and only if there is a nontrivial relation among the generators in S. To show this, suppose $x_0, x_1,\ldots, x_n = x_0$ is a cycle of $G(H, S)$. Then, there are $a_i \in S$, $1 \leq i \leq n$, such that $x_{i-1}a_i^{\varepsilon_i} = x_i$, where $\varepsilon_i \in \{1,-1\}$. Hence, $x_n = x_{n-1}a_n^{\varepsilon_n} = x_{n-2}a_{n-1}^{\varepsilon_{n-1}} a_n^{\varepsilon_n} = \ldots = x_0 a_1^{\varepsilon_1}a_2^{\varepsilon_2}\ldots a_n^{\varepsilon_n}$, that is, the identity $1 = a_1^{\varepsilon_1}a_2^{\varepsilon_2}\ldots a_n^{\varepsilon_n}$. If this were a trivial relation, then there would exist an integer i, $1 \leq i \leq n$, such that $a_i = a_{i+1}$ and $\varepsilon_i = -\varepsilon_{i+1}$. However, this implies that $x_{i-1} = x_{i+1}$, a contradiction. Similarly, if $a_1^{\varepsilon_1} a_2^{\varepsilon_2} ... a_n^{\varepsilon_n} = 1$ is a nontrivial relation, then x_0, x_1,..., x_{n-1}, x_n, where $x_i = x_{i-1} a_i^{\varepsilon_i}$, $1 \leq i \leq n$, and $x_0 = x_n$, is a closed trial in $G(H, S)$, which must contain a cycle.

Suppose now that H is a free group, S a minimal set of generators of H, and J a subgroup of H. Since there is no nontrivial relation on the elements of S, $G(H, S)$ does not contain a cycle. Also, from the Corollary above, $G(H, S)$ is contractible onto $G(J, T)$ for some set T of generators of J. Because any contraction of a cycle-free graph is again cycle free, $G(J, T)$ must be cycle free, and, thus, there is no nontrivial relation on the elements of T. Hence, J must be a free group, freely generated by T. ■

4. The SNP Assembly Problem

In computational biochemistry there are many situations where we wish to resolve conflicts between sequences in a sample by excluding some of the sequences. Of course, exactly what constitutes a conflict must be precisely defined in the biochemical context.

We define a *conflict graph* where the vertices represent the sequences in the sample and there is an edge between two vertices if and only if there is a conflict between the corresponding sequences. The aim is to remove the fewest possible sequences that will eliminate all conflicts.

11

Recall that given a simple graph G, a *vertex cover* C is a subset of the vertices such that every edge has at least one end in C. Thus, the aim is to find a minimum vertex cover in the conflict graph G. (in general, this is known to be a NP-complete problem [13]). We look at a specific example of the SNP assembly problem given in [15] and show how to solve this problem using the vertex cover algorithm [6].

A *Single Nucleotide Polymorphism* (*SNP*, pronounced "snip") [15] is a single base mutation in *DNA*. It is known that *SNP*s are the most common source of genetic polymorphism in the human genome (about 90% of all human *DNA* polymorphisms).

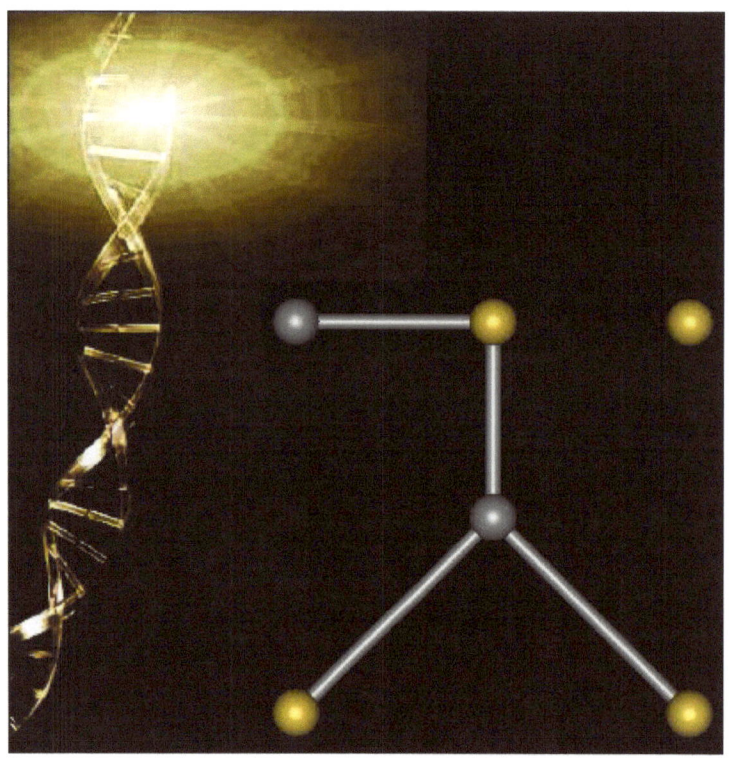

Figure 4.1. *The DNA double helix and SNP assembly problem*

The *SNP Assembly Problem* [15] is defined as follows. A *SNP assembly* is a triple (S, F, R) where $S = \{s_1, \ldots, s_n\}$ is a set of n SNPs, $F = \{f_1, \ldots, f_m\}$ is a set of m fragments and R is a relation $R: S \times F \to \{0, A, B\}$ indicating whether a *SNP* $s_i \in S$ does not occur on a fragment $f_j \in F$ (marked by 0) or if occurring, the non-zero value of s_i (A or B). Two *SNPs* s_i and s_j are defined to be in *conflict* when there exist two fragments f_k and f_l such that exactly three of $R(s_i, f_k)$, $R(s_i, f_l)$, $R(s_j, f_k)$, $R(s_j, f_l)$ have the same non-zero value and exactly one has the opposing non-zero value. The problem is to remove the fewest possible *SNPs* that will eliminate all conflicts. The following example from [15] is shown in the table below. Note that the relation R is only defined for a subset of $S \times F$ obtained from experimental values.

R	f_1	f_2	f_3	f_4	f_5
s_1	A	B			B
s_2	B	A	A	A	0
s_3	0	0	B	B	A
s_4	A	0	A	0	B
s_5	A	B	B	B	A
s_6	B		A	A	0

Note, for instance, that s_1 and s_5 are in conflict because $R(s_1, f_2) = B$, $R(s_1, f_5) = B$, $R(s_5, f_2) = B$, $R(s_5, f_5) = A$. Again, s_4 and s_6 are in conflict because $R(s_4, f_1) = A$, $R(s_4, f_3) = A$, $R(s_6, f_1) = B$, $R(s_6, f_3) = A$. Similarly, all pairs of conflicting *SNP*s are easily determined from the table. The conflict graph G corresponding to this *SNP* assembly problem is shown below in figure 4.2.

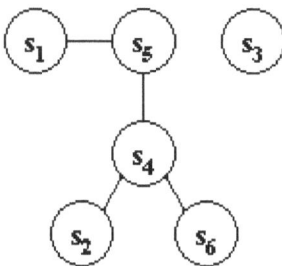

Figure 4.2. The conflict graph G

We now use the vertex cover algorithm [6] to find minimal vertex covers in the conflict graph G. The input is the number of vertices 6, followed by the adjacency matrix of G shown below in figure 4.3. The entry in row i and column j of the adjacency matrix is 1 if the vertices s_i and s_j have an edge in the conflict graph and 0 otherwise.

```
0 0 0 0 1 0
0 0 0 1 0 0
0 0 0 0 0 0
0 1 0 0 1 1
1 0 0 1 0 0
0 0 0 1 0 0
```

Figure 4.3. The input for the vertex cover algorithm

The vertex cover program [6] finds two distinct minimum vertex covers, shown in figure 4.4.

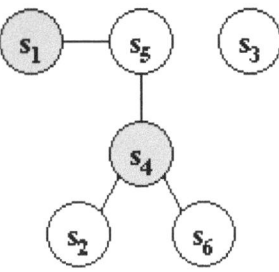

Minimum Vertex Cover: s_1, s_4

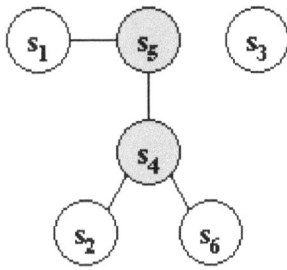

Minimum Vertex Cover: s_4, s_5

Figure 4.4. *The output of the vertex cover algorithm*

Thus, either removing s_1, s_4 or removing s_4, s_5 solves the given *SNP* assembly problem. ∎

5. Computer Network Security

A team of computer scientists led by Eric Filiol [11] at the Virology and Cryptology Lab, ESAT, and the French Navy, ESCANSIC, have recently used the vertex cover algorithm [6] to simulate the propagation of stealth worms on large computer networks and design optimal strategies for protecting the network against such virus attacks in real-time.

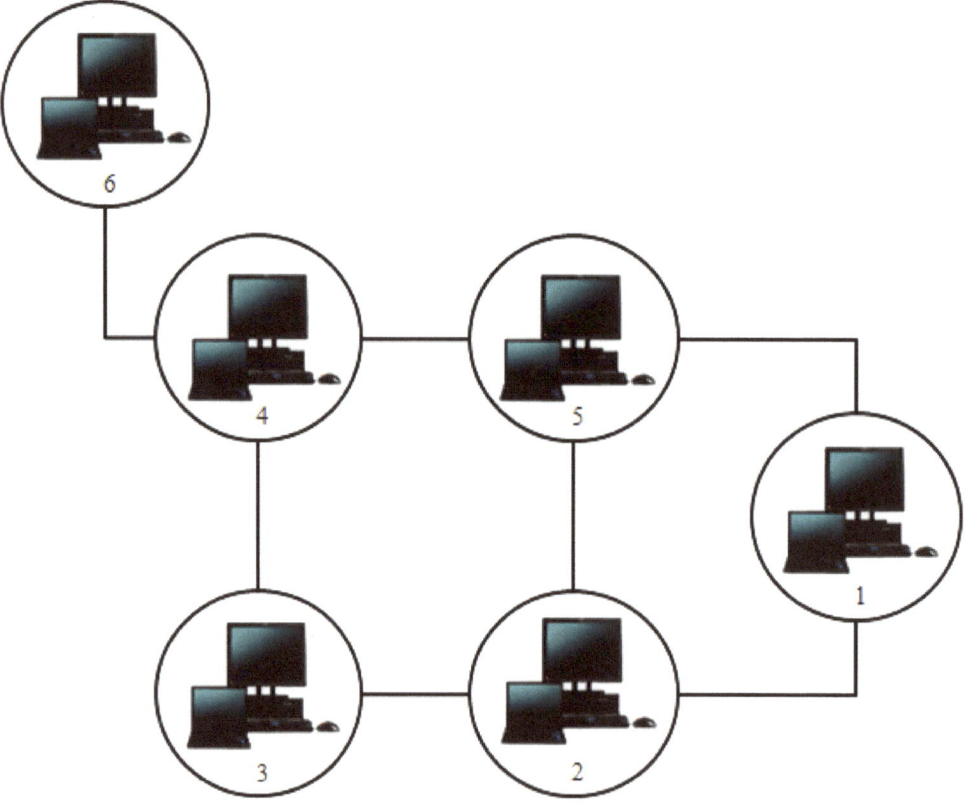

Figure 5.1. *The set {2, 4, 5} is a minimum vertex cover in this computer network*

The simulation was carried out on a large internet-like virtual network and showed that that the combinatorial topology of routing may have a huge impact on the worm propagation and thus some servers play a more essential and significant role than others. The real-time capability to identify them is essential to greatly hinder worm propagation. The idea is to find a minimum vertex cover in the graph whose vertices are the routing servers and whose edges are the (possibly dynamic) connections between routing servers. This is an optimal solution for worm propagation and an optimal solution for designing the network defense strategy. Figure 5.1 above shows a simple computer network and a corresponding minimum vertex cover $\{2, 4, 5\}$.

6. The Timetabling Problem

In a college there are m professors $x_1, x_2, \ldots, x_m$ and n subjects $y_1, y_2, \ldots, y_n$ to be taught. Given that professor x_i is required (and able) to teach subject y_j for p_{ij} periods ($p = [p_{ij}]$ is called the *teaching requirement matrix*), the college administration wishes to make a timetable using the minimum possible number of periods. This is known as the *timetabling problem* [4] and can be solved using the following strategy. Construct a bipartite multigraph G with vertices $x_1, x_2, \ldots, x_m, y_1, y_2, \ldots, y_n$ such that vertices x_i and y_j are connected by p_{ij} edges. We presume that in any one period each professor can teach at most one subject and that each subject can be taught by at most one professor. Consider, first, a single period. The timetable for this single period

17

corresponds to a matching in the graph and, conversely, each matching corresponds to a possible assignment of professors to subjects taught during this period. Thus, the solution to the timetabling problem consists of partitioning the edges of G into the minimum number of matchings. Equivalently, we must properly color the edges of G with the minimum number of colors. We shall show yet another way of solving the problem using the vertex coloring algorithm [7].

Recall that the *line graph* $L(G)$ of G has as vertices the edges of G and two vertices in $L(G)$ are connected by an edge if and only if the corresponding edges in G have a vertex in common. The line graph $L(G)$ is a simple graph and a proper vertex coloring of $L(G)$ yields a proper edge coloring of G using the same number of colors. Thus, to solve the timetabling problem, it suffices to find a minimum proper vertex coloring of $L(G)$ using [7]. We demonstrate the solution with a small example.

Suppose there are four professors x_1, x_2, x_3, x_4 and five subjects y_1, y_2, y_3, y_4, y_5 to be taught [4]. The teaching requirement matrix $p = [p_{ij}]$ is given below in figure 6.1.

p	y_1	y_2	y_3	y_4	y_5
x_1	2	0	1	1	0
x_2	0	1	0	1	0
x_3	0	1	1	1	0
x_4	0	0	0	1	1

Figure 6.1. The teaching requirement matrix

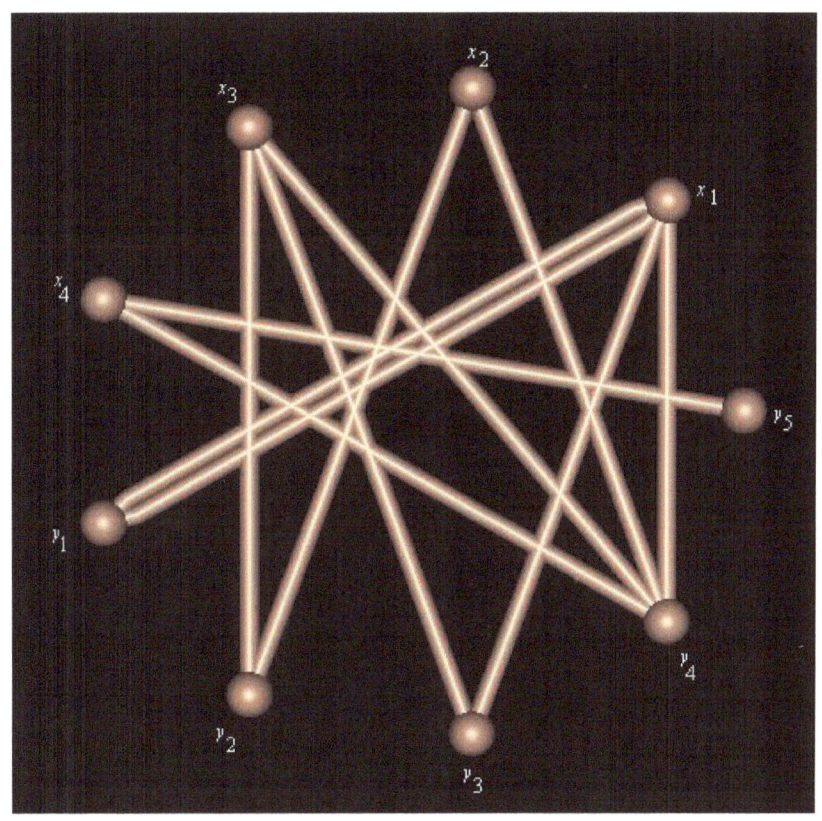

Figure 6.2. *The bipartite multigraph G*

We first construct the bipartite multigraph G shown above in figure 6.2. Next, we construct the line graph $L(G)$. The adjacency matrix of $L(G)$ is given below.

```
0 1 1 1 0 0 0 0 0 0 0
1 0 1 1 0 0 0 0 0 0 0
1 1 0 1 0 0 0 1 0 0 0
1 1 1 0 0 1 0 0 1 1 0
0 0 0 0 0 1 1 0 0 0 0
0 0 0 1 1 0 0 0 1 1 0
0 0 0 0 1 0 0 1 1 0 0
0 0 1 0 0 0 1 0 1 0 0
0 0 0 1 0 1 1 1 0 1 0
0 0 0 1 0 1 0 0 1 0 1
0 0 0 0 0 0 0 0 0 1 0
```

Now, we use the vertex coloring algorithm [7] to find a minimum proper 4-coloring of the vertices of $L(G)$.

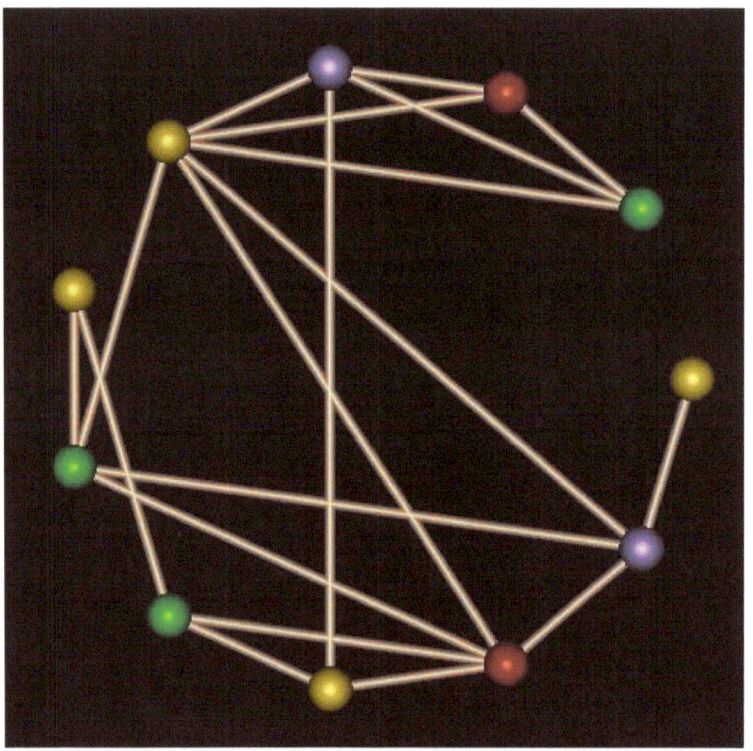

Figure 6.3. *A minimum proper 4-coloring of the vertices of* $L(G)$

Vertex Coloring: (1 , green) (2 , red) (3 , blue) (4 , yellow) (5 , yellow) (6 , green) (7 , green) (8 , yellow) (9 , red) (10 , blue) (11 , yellow). This, in turn, yields a minimum proper edge 4-coloring of the bipartite multigraph G:

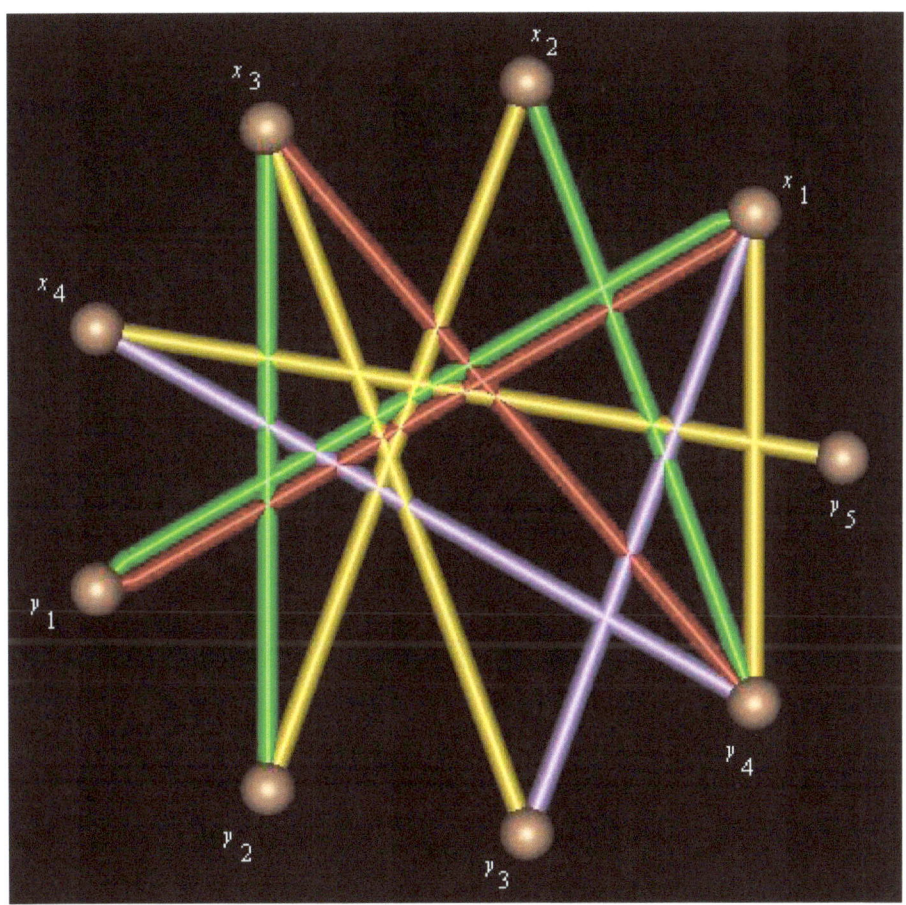

Figure 6.4. *A minimum proper 4-coloring of the edges of G*

Edge Coloring: ($\{x_1,y_1\}$, green) ($\{x_1,y_1\}$, red) ($\{x_1,y_3\}$, blue) ($\{x_1,y_4\}$, yellow) ($\{x_2,y_2\}$, yellow) ($\{x_2,y_4\}$, green) ($\{x_3,y_2\}$, green)

21

({x_3,y_3} , yellow) ({x_3,y_4} , red) ({x_4,y_4} , blue) ({x_4,y_5} , yellow).
Interpret the colors green, red, blue, yellow as periods 1, 2, 3, 4 respectively.
Then, from the edge coloring of G, we obtain a solution of the given
timetabling problem as shown below in figure 6.5.

	1	2	3	4
x_1	y_1	y_1	y_3	y_4
x_2	y_4			y_2
x_3	y_2	y_4		y_3
x_4			y_4	y_5

Figure 6.5. The timetable

7. Map Coloring and GSM Mobile Phone Networks

Given a map drawn on the plane or the surface of a sphere, the famous four
color theorem asserts that it is always possible to properly color the regions
of the map such that no two adjacent regions are assigned the same color,
using at most four distinct colors [8] [18] [1]. For any given map, we can
construct its dual graph as follows. Put a vertex inside each region of the map
and connect two distinct vertices by an edge if and only if their respective
regions share a whole segment of their boundaries in common. Then, a

proper vertex coloring of the dual graph yields a proper coloring of the regions of the original map.

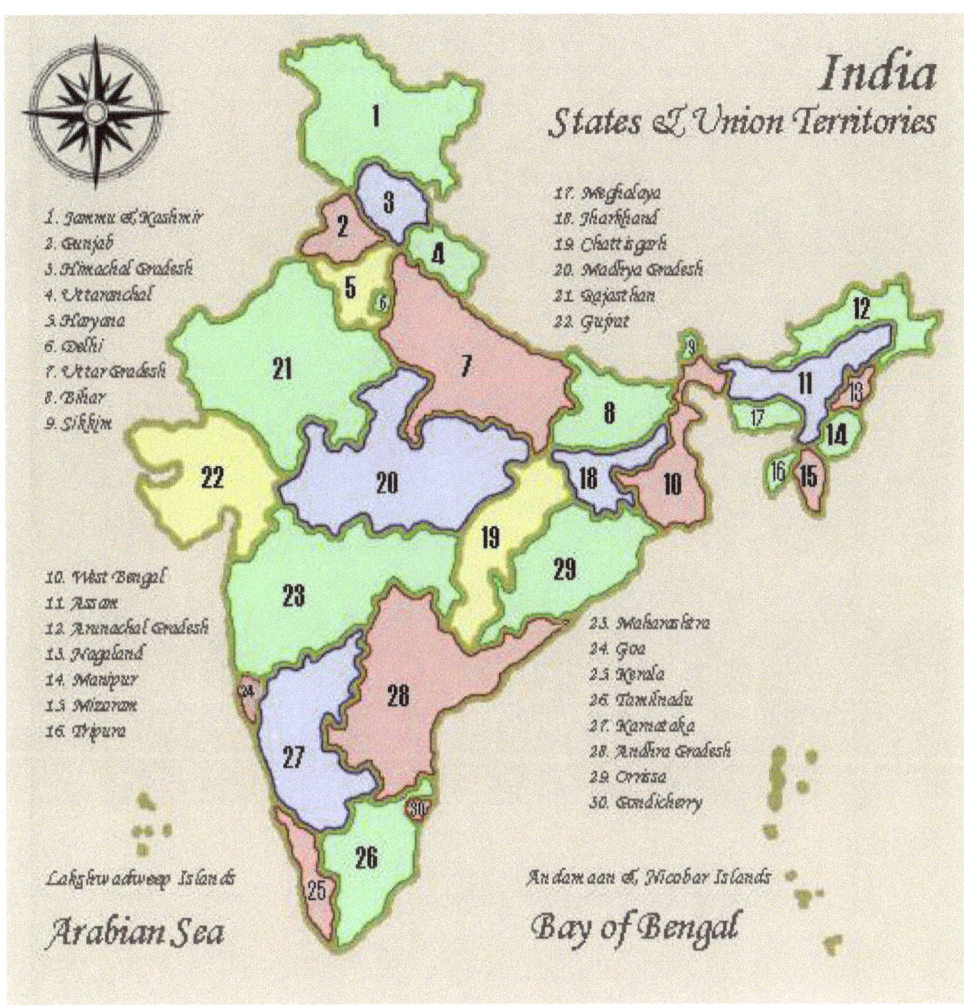

Figure 7.1. The map of India

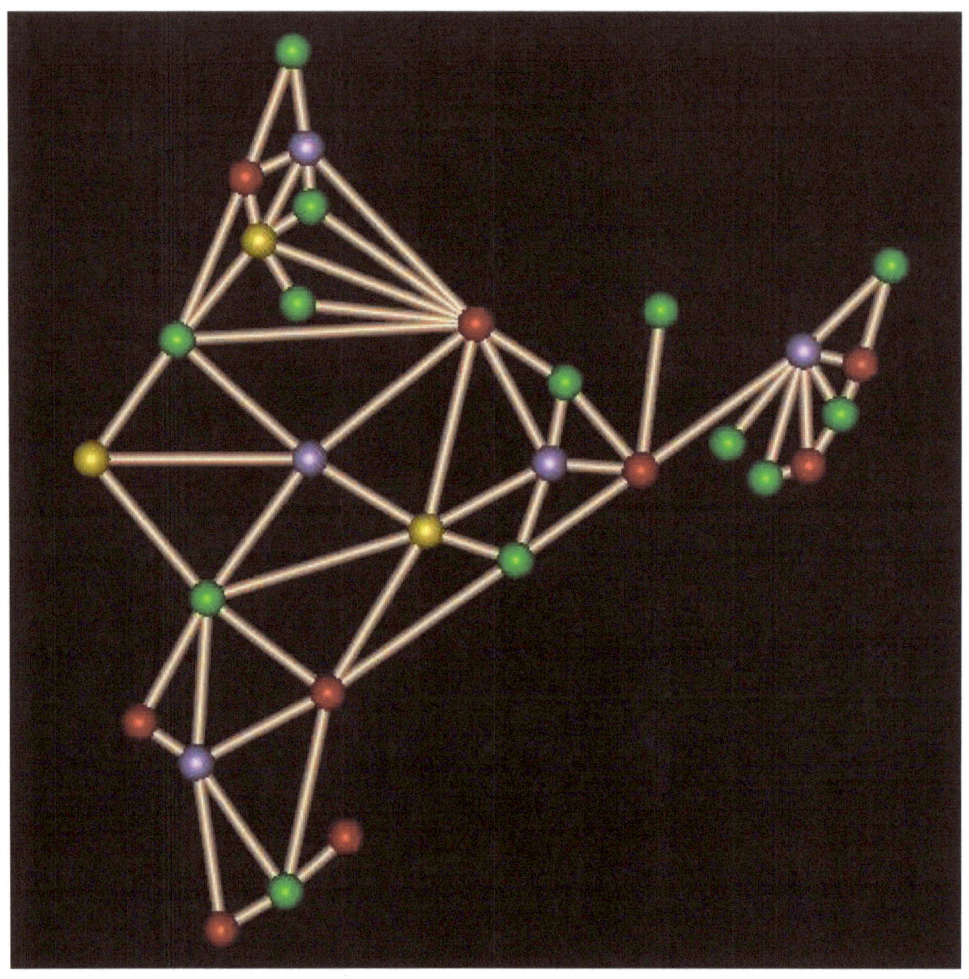

Figure 7.2. *The dual graph of the map of India*

We use the vertex coloring algorithm [7] to find a proper coloring of the map of India with four colors, see figures 7.1 and 7.2 above.

The Groupe Spécial Mobile (GSM) was created in 1982 to provide a standard for a mobile telephone system. The first GSM network was launched in 1991 by Radiolinja in Finland with joint technical infrastructure maintenance from Ericsson. Today, GSM is the most popular standard for mobile phones in the world, used by over 2 billion people across more than 212 countries. GSM is a cellular network with its entire geographical range divided into hexagonal cells. Each cell has a communication tower which connects with mobile phones within the cell. All mobile phones connect to the GSM network by searching for cells in the immediate vicinity. GSM networks operate in only four different frequency ranges. The reason why only four different frequencies suffice is clear: the map of the cellular regions can be properly colored by using only four different colors! So, the vertex coloring algorithm may be used for assigning at most four different frequencies for any GSM mobile phone network, see figure 7.2 below.

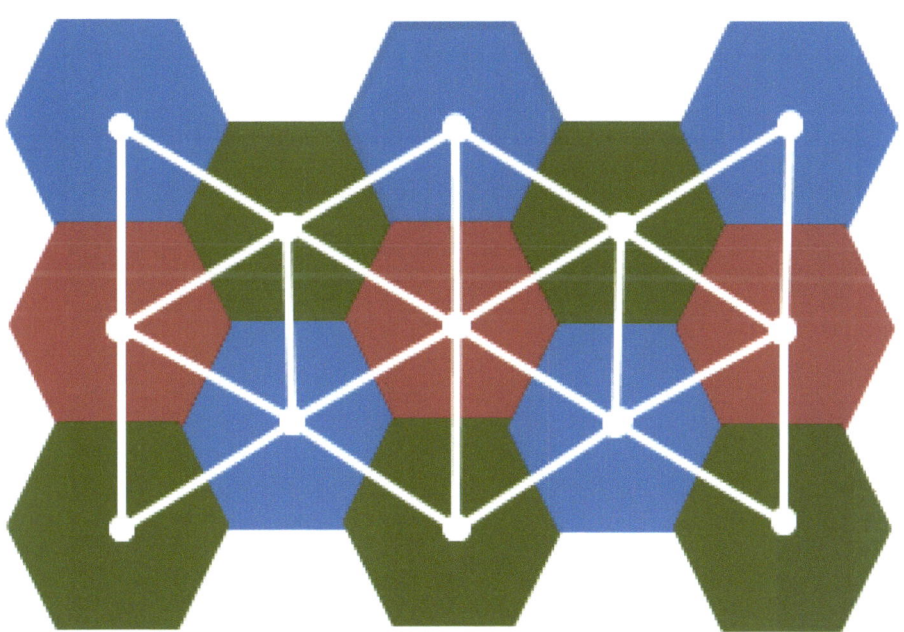

Figure 7.3. The cells of a GSM mobile phone network

8. Knight's Tours

In 840 A.D., al-Adli [17], a renowned shatranj (chess) player of Baghdad is said to have discovered the first *re-entrant knight's tour*, a sequence of moves that takes the knight to each square on an 8×8 chessboard exactly once, returning to the original square. Many other re-entrant knight's tours were subsequently discovered but Euler [10] was the first mathematician to do a systematic analysis in 1766, not only for the 8×8 chessboard, but for re-entrant knight's tours on the general $n \times n$ chessboard. Given an $n \times n$ chessboard, define a knight's graph with a vertex corresponding to each square of the chessboard and an edge connecting vertex i with vertex j if and only if there is a legal knight's move from the square corresponding to vertex i to the square corresponding to vertex j. Thus, a re-entrant knight's tour on the chessboard corresponds to a Hamiltonian circuit in the knight's graph. The Hamiltonian circuit algorithm [9] [13] has been used to find re-entrant knights tours on chessboards of various dimensions.

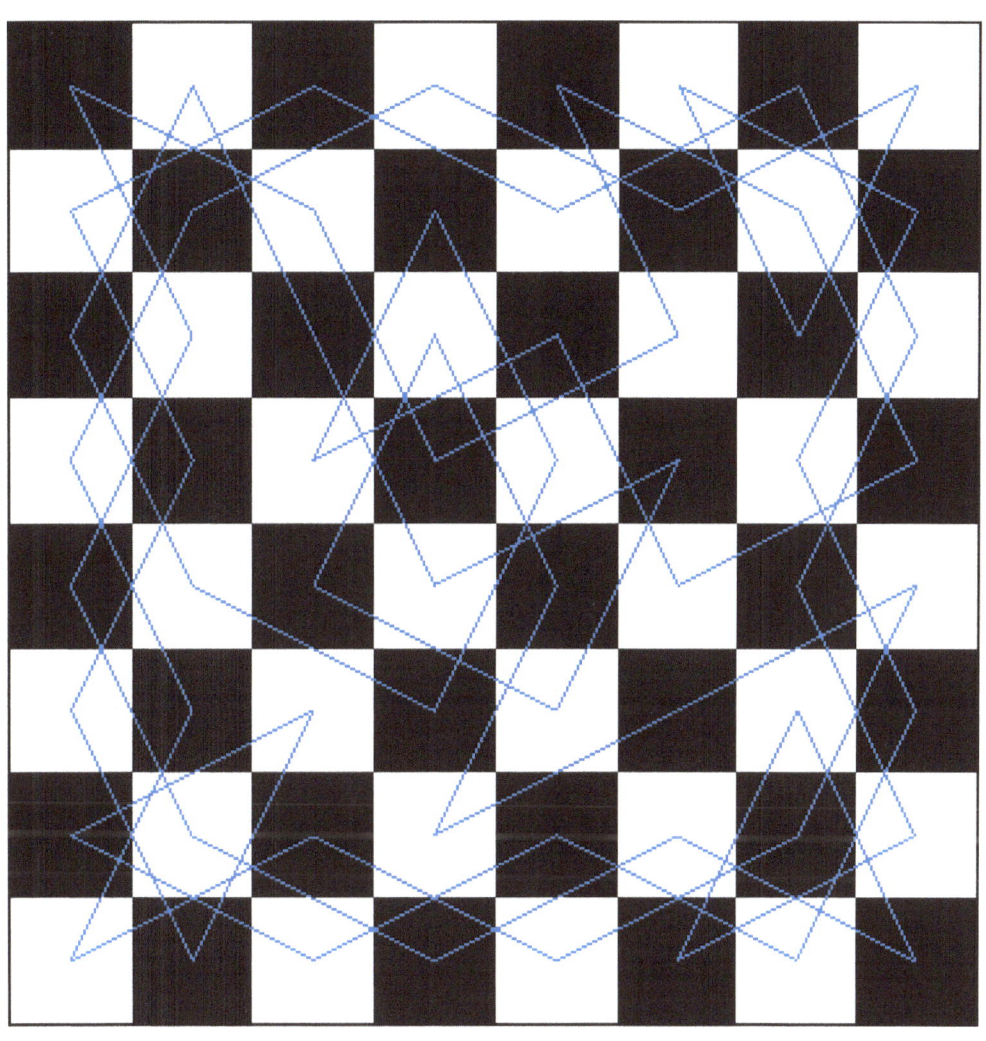

Figure 8.1. *A re-entrant knight's tour on the 8×8 chessboard*

References

[1]   K. Appel and W. Haken, *Every Planar Map is Four Colorable*, Bull. Amer. Math. Soc. 82 (1976) 711-712.

[2]   L. Babai, *Some applications of graph contractions*, J. Graph Theory, Vol. 1 (1977) 125-130.

[3]   E. Bertram and P. Horak, *Some applications of graph theory to other parts of mathematics*, The Mathematical Intelligencer (Springer-Verlag, New York) (1999) 6-11.

[4]   J.A. Bondy and U.S.R. Murty, *Graph Theory with Applications*, 1976, Elsevier Science Publishing Company Inc.

[5]   L. Caccetta and K. Vijayan, *Applications of graph theory*, Fourteenth Australasian Conference on Combinatorial Mathematics and Computing (Dunedin, 1986), Ars. Combin., Vol. 23 (1987) 21-77.

[6]   Ashay Dharwadker, *The Vertex Cover Algorithm*, 2006, **http://www.dharwadker.org/vertex_cover**

[7]   Ashay Dharwadker, *The Vertex Coloring Algorithm*, 2006, **http://www.dharwadker.org/vertex_coloring**

[8]   Ashay Dharwadker, *A New Proof of The Four Colour Theorem*, 2000, **http://www.dharwadker.org**

[9]   Ashay Dharwadker, *A New Algorithm for finding Hamiltonian Circuits*, 2004, **http://www.dharwadker.org/hamilton**

[10] L. Euler, *Solution d'une question curieuse qui ne paroit soumise a aucune analyse*, Mémoires de l'Académie Royale des Sciences et Belles Lettres de Berlin, Année 1759 15, 310-337, 1766.

[11] Eric Filiol, Edouard Franc, Alessandro Gubbioli, Benoit Moquet and Guillaume Roblot, *Combinatorial Optimisation of Worm Propagation on an Unknown Network*, Proc. World Acad. Science, Engineering and Technology, Vol 23, August 2007, **http://www.waset.org/journals/waset/v34/v34-8.pdf**

[12] K. Heinrich and P. Horak, *Euler's theorem*, Am. Math. Monthly, Vol. 101 (1994) 260.

[13] R.M. Karp, *Reducibility among combinatorial problems*, Complexity of Computer Computations, Plenum Press, 1972.

[14] D. König, *Theorie der endlichen und unendlichen graphen*, Akademische Verlagsgesllschaft, Leipzing (1936), reprinted by Chelsea, New York (1950).

[15] G. Lancia, V. Afna, S. Istrail, L. Lippert, and R. Schwartz, *SNPs Problems, Complexity and Algorithms*, ESA 2002, LNCS 2161, pp. 182-193, 2001. Springer-Verlag 2001.

[16] L. Lovasz, L. Pyber, D. J. A. Welsh and G. M. Ziegler, *Combinatorics in pure mathematics*, in Handbook of Combinatorics, Elsevier Sciences B. V., Amsterdam (1996).

[17] H. J. R. Murray, *A History of Chess*, Oxford University Press, 1913.

[18] Shariefuddin Pirzada and Ashay Dharwadker, *Graph Theory*, Orient Longman and Universities Press of India, 2007.

[19] F. S. Roberts, *Graph theory and its applications to the problems of society*, CBMS-NSF Monograph 29, SIAM Publications, Philadelphia (1978).

[20] J. P. Serre, *Groupes Discretes*, Extrait de I'Annuaire du College de France, Paris (1970).

[21] R. Thomas, *A combinatorial construction of a non-measurable set*, Am. Math. Monthly 92 (1985) 421-422.